KAGEKI SHOJO!!

story & art by
Kumiko Saiki

7

Characters
The Kouka Theater Troupe
A COMPANY OF UNMARRIED FEMALE ACTRESSES ESTABLISHED IN THE TAISHO ERA. OUR STORY FOLLOWS THE YOUNG WOMEN OF THE 100TH CLASS OF ASPIRING ACTRESSES AT THE KOUKA SCHOOL OF MUSICAL AND THEATRICAL ARTS, WHERE THEY WILL BE TRAINED TO BECOME THE NEXT GENERATION OF KOUKA PERFORMERS.
Narata Ai (16)
Former member of the extremely popular idol group JPX48.
Watanabe Sarasa (16)
Ditzy girl standing tall at a height of 178 cm. Her dream: "To be Lady Oscar!"
First-Year Students
Studying music, dance, and theater
Hoshino Kaoru (18)
Third-generation Kouka thoroughbred.
Sugimoto Sawa (16)
Class rep. Top of the class in grades. Huge Kouka nerd.
Sawada Chika (16)
Sawada Chiaki (16)
Last year, Chika passed, but Chiaki didn't, so they waited a year and tried again.
Yamada Ayako (16)
Best singer in the class. Worried about her weight.

Second-Year Students
Mentors and rivals to the first-years
Takei
Second-Year Class Rep.
Nakayama Risa
Sarasa's mentor. Half-Latina.
Nojima Hijiri
Ai's mentor. Huge JPX48 fan.
Otokoyaku Top Stars
Spring Troupe
Asahina Ryuu
Summer Troupe
Shiina Reo
Autumn Troupe
Mitsuki Keito
Winter Troupe
Satomi Sei
Kouka School Teachers
Andou Mamoru
Acting teacher.
Narata Taichi
Ai's uncle. Ballet teacher.
Kabuki Actors
Shirakawa Kaou
15th Generation Kaou and National Treasure.Might be related to Sarasa...?!
Shirakawa Kouzaburou
Kabuki musumeyaku. Sexy kabuki star.
Shirakawa Akiya
Sarasa's childhood friend and boyfriend (?). Will likely become the 16th Shirakawa Kaou.

CONTENTS

No.
Date

Kouka Troupe Terms to Know!★

Kouka Theater Troupe
Founded a hundred years ago as a theater troupe comprised of young, unmarried women. Split into four troupes (Spring/Summer/Autumn/Winter). Main theater is in Kobe.

Otokoyaku / Musumeyaku
Designated roles for the gender of characters actresses play. Otokoyaku actresses play male or masculine characters, while musumeyaku actresses play female or feminine characters.

Top Star
The actress who heads her troupe. Each troupe has an otokoyaku top star and musumeyaku top star. Top stars appear in every major production.

Kouka School of Musical and Theatrical Arts
Two-year prep school where the next stars of the Kouka Troupe are forged. Girls can apply anytime between 9th and 12th grade!

First-Year Students vs. Second-Year Students
While first-year students primarily focus on their studies, second-year students are tasked both with their studies and mentoring the first-year students, as well as managing their cleaning schedule and helping them with lifestyle adjustments.

21

IN THE BROADWAY MUSICAL AMADEUS...
TALENTED PALACE COMPOSER ANTONIO SALIERI...
GROWS JEALOUS OF COMPOSER WOLFGANG MOZART'S INHERENT MUSICAL GENIUS, AND THROUGH HIS ATTEMPTS TO RUIN MOZART'S REPUTATION AND CAREER, DESTROYS HIS OWN.
AT LEAST, THAT WAS THE STORY IN THE MUSICAL.

ROMEO: SHIROTA MINAMI.
JULIET: YAMADA AYAKO.
NURSE: HAMADA ASAMI.
TYBALT: WATANABE SARASA.
YOU FOUR ARE OUR CAST FOR THE *ROMEO AND JULIET* READING AT THE CULTURE FESTIVAL.
YES, SIR!
BE SURE TO PRACTICE ON YOUR OWN AND AS A GROUP.
YES, SIR!!

OH, AND I'LL COME WATCH YOU REHEARSE THREE TIMES BEFORE DRESS REHEARSAL.
THREE TIMES?
ONLY THREE TIMES?
THE STARS OF THE CULTURE FESTIVAL ARE THE SECOND-YEARS.
YOU GUYS ARE A FUN BONUS.
AND YOUR LINES ARE MOSTLY SOLILOQUIES, WITH NO INTERACTION BETWEEN YOU. AS LONG AS YOU CAN SAY YOUR LINES WITHOUT FLUBBING THEM, YOU'RE GOOD.
OKAY.
BREAK A LEG!
WELL, NOW THIS DOESN'T SEEM LIKE THAT BIG OF A DEAL.
YUP.
BUT LET'S GIVE IT OUR ALL, ANYWAYS!!
YEAH!
LET'S DO THIS!

I DON'T *CARE* IF HE SAID YOU'RE JUST A "FUN BONUS"!
YOU'RE REPRESENTING US OUT THERE, SO YOU *BETTER* BE AMAZING!
ROGER THAT!!
WHO WOULD HAVE GUESSED THE TWO WORST STUDENTS IN OUR CLASS WOULD GET LEAD ROLES?
IT'S NOT UNHEARD OF.
THERE HAVE EVEN BEEN TOP STARS WHO SCORED THE LOWEST ON THE ENTRANCE EXAM FOR THEIR CLASS.
WOW, REALLY?!
THERE'S ALSO AN URBAN LEGEND THAT SAYS THAT ACTRESSES WHO MESS UP IN THE PARADE OR FALL OFF THE GRAND STAIRS WILL BECOME A TOP STAR ONE DAY.

THERE'S THE ONE ABOUT STANDING UNDER THE CHERRY TREE BEFORE YOU GET IN MEANS YOU'LL NEVER BE A TOP STAR.
OR THE ONE WHERE STUDENTS WHO THINK PEOPLE WITH GOOD GRADES WILL BE TOP STARS...
ALL DO TERRIBLY ONCE THEY'RE IN THE ACTUAL TROUPE.
THERE'S ALSO AN URBAN LEGEND ABOUT A MONSTER THAT LIVES UNDER THE STAGE.
SO REALLY, ANYTHING GOES IN THE WORLD OF KOUKA.
SAWA!
THE SECOND-YEAR CLASS REP WANTS YOU TO COME TO THE JAPANESE ROOM WHEN YOU'RE DONE EATING!
GOT IT. I'LL HEAD THERE NOW.
IT'S PROBABLY ABOUT THE CULTURE FESTIVAL.
......
nom-nom

SO, UH!
ARE YOUR FAMILIES GOING TO COME WATCH THE CULTURE FESTIVAL SHOW?
MY FAMILY GOT SO EXCITED AND SAID THEY'D *ALL* COME.
AND MY SISTER BURST INTO TEARS AGAIN.
OUR MOM'S COMING, BUT OUR DAD HAS TO WORK.
MY MOM AND GRANDMOTHER ARE BOTH COMING.
NO WAY!
SPRING'S SNOW WHITE IS COMING?!
OH MY GOSH, OH MY GOSH!!
OUR MOM'S GOING TO *FLIP!*
MY GRANDPA AND KEIKO FROM THE SOBA SHOP AND SHOU-CHAN THE RICKSHAW DRIVER ARE ALL GONNA COME!
THERE SHE GOES!
YEP, THERE SHE GOES, RATTLING OFF THOSE NAMES LIKE WE KNOW THEM!
WHO ARE THEY, ANYWAY?!
MY FRIENDS FROM THE NEIGHBOR-HOOD!
HOW 'BOUT YOU, NARACCHI?

NO ONE...
IS COMING.
AWW, YOUR MOM'S NOT GONNA COME?
SHE'S NARATA KIMIKO, THE ACTRESS, RIGHT?
I WATCHED AN OLD MOVIE NOT TOO LONG AGO AND SAW HER IN IT AS HOSOKAWA GRACIA!
OH, YEAH! SHE WAS SOOOO GORGEOUS IN IT! ♡
IT'D JUST BE ANNOYING IF SHE CAME.
SO NO FAMILY'S COMING.
NO FAMILY?

WELL, MY UNCLE WILL--
NARATA-SENSEI DOESN'T COUNT!
HM?
AND WHEN THE CULTURE FESTIVAL'S OVER, WE'LL FINALLY BE BEAUTIFUL BIG SISTERS OURSELVES! ♡
CAN YOU STOP CALLING THE SECOND-YEARS THAT?
YAAAY!!
WE'LL FINALLY BE FREE FROM MORNING CLEANING!
AND WE CAN EAT LUNCH IN THE CLASSROOM INSTEAD OF THE LOCKER ROOM!
First-Years

THERE ARE TWO MONTHS LEFT UNTIL THE CULTURE FESTIVAL.
US SECOND-YEARS WILL BE SHIFTING FROM CLASSES...
TO REHEARSALS FOR THE FESTIVAL.
WHILE THE CULTURE FESTIVAL WILL BE SEEN BY OUR FAMILIES AND MEMBERS OF THE GENERAL PUBLIC...
THE FESTIVAL IS ALSO A ONCE-IN-A-LIFETIME OPPORTUNITY FOR US TO BOND AS CLASSMATES.
IT'S JUST AS IMPORTANT TO US AS OUR DEBUT PERFORMANCES!
AS SUCH, WE'LL BE COUNTING ON YOU FIRST-YEARS TO HELP US OUT FOR THE NEXT TWO MONTHS.

THE BEST THING YOU CAN DO IS TRY NOT TO TRIP US UP. ♡
ESPECIALLY SINCE YOU GET TO HAVE A LITTLE SKIT IN *OUR SHOW!*
YOU'LL BE PLAYING ROMEO, SHIROTA-SAN?
YES, MA'AM!
I'LL DO EVERYTHING I CAN TO AVOID CAUSING PROBLEMS FOR YOUR CLASS!!
AND TYBALT WILL BE PLAYED BY WATANABE-SAN, CORRECT?

I SURE HOPE WATANABE-SAN IS GOING TO BE OKAY!
ESPECIALLY AFTER SHE *FELL* AT THE SPORTS FESTIVAL!
SHE DOESN'T REALLY FIT THE DARK IMAGE OF TYBALT, DOES SHE?
WATANABE-SAN'S PERFORMANCE AS TYBALT...

WAS INCREDIBLE.
I DIDN'T COME HERE TO PROTEST THE CASTING.

SENSEI.
TELL ME, WHAT WAS IT THAT I WAS LACKING?
WHAT YOU LACKED, HUH?
WELL, YOU DON'T HAVE EXPERIENCE YET, SO YOU'RE LACKING ALL SORTS OF SKILLS.
AND IT'S NOT JUST YOU. IT'S EVERYONE IN YOUR CLASS.

......
DO YOU KNOW THE MUSICAL AMADEUS, SIR?
THE BROADWAY SHOW?
SURE. THE MOVIE WAS PRETTY GOOD, TOO.
THAT'S THE ONE ABOUT MOZART AND SALIERI...
AND THE CLASH BETWEEN GENIUS AND HARD WORK, THE SUSPENSEFUL STORY OF TWO COMPOSERS.
SALIERI COULD NOT COMPREHEND WHY GOD HAD CHOSEN TO BLESS MOZART WITH THE TALENT FOR COMPOSING.
FOR WHILE MOZART LOVED MUSIC DEARLY, HE WAS FOOLISH AND UNCOUTH--YET THE POPULACE COULD ONLY SEE HIM FOR HIS MUSIC AND THE TALENT THAT CREATED IT.
SALIERI FORSOOK HIS FAITH IN GOD AND DEVOTED HIMSELF TO DISPARAGING MOZART'S REPUTATION.
BOTH CHARACTERS ARE EXTREMELY COMPELLING.

I HATE THAT MUSICAL.
WHEN I FIRST SAW IT...
I IDENTIFIED WITH SALIERI, AND IT SCARED ME.
I CAN'T HELP BUT FEEL THAT NO MATTER HOW SKILLED I AM...
I'LL NEVER HOLD A CANDLE TO PRODIGIES LIKE WATANABE-SAN.
I WORRY THAT NO MATTER HOW HARD I WORK, I'LL NEVER OUTMATCH HER.
AND PART OF ME ADMIRES HER EFFORTLESS ABILITY...

BUT I KNOW THAT IF SHE WASN'T HERE, I WOULDN'T FEEL SO INSECURE.
THAT IF SHE HADN'T TRIED OUT FOR TYBALT, IT COULD HAVE BEEN *ME*.
I CAN'T HELP BUT THINK THESE THINGS, AND I HATE MYSELF FOR IT.
BUT, *UH*...
NOW THAT I'M SAYING ALL THIS OUT LOUD...
I REALIZE IT JUST SOUNDS LIKE I'M BEING PETTY.
I'M SORRY FOR BOTHERING YOU ABOUT THIS!
IT'S FINE, DON'T WORRY.
THIS IS WAY BETTER THAN YOU SITTING HERE AND SOBBING.

DID PEOPLE REALLY SIT HERE AND CRY?
SURE DID.
WOULD IT CHANGE ANYTHING IF I CRIED?
IT MIGHT MAKE YOU FEEL BETTER.
BUT PLEASE DON'T. I'M NOT GOOD WITH TEARS.
I WON'T CRY.
I SAVE MY TEARS FOR THE THEATER.
I THOUGHT NOT.
OKAY, SO...
MOST "TALENTED" PEOPLE ARE LIKE SALIERI.
ACTING IS ALL ABOUT TEAMWORK, AND THE KOUKA TROUPE VALUES INDIVIDUALITY.
WE DON'T WANT A GROUP OF JUST MOZARTS.
IF *EVERYONE* IN THE TROUPE WERE UNPRECEDENTED PRODIGIES, IT'D BE TOO ONE-NOTE.

YOU SHOULD ALSO KNOW THAT THE VOTES WERE EVENLY SPLIT BETWEEN YOU AND WATANABE.
SO THE TEACHERS VOTED AMONGST OURSELVES.
AND WE WERE *STILL* SPLIT ON YOU TWO.
OHGI-SENSEI CAME LATE, SO WHEN SHE ARRIVED...
HUH? WHAT?!
WE HAD HER CAST THE TIE-BREAKER.
I see where you're coming from, now.

TYBALT
Sugimoto Sawa: 17
Watanabe Sarasa: 17
Right then!
Tybalt should be played by Watanabe Sarasa!

YOU MISSED IT BY A PAPER-THIN MARGIN.
SO SURE, YOU MIGHT HAVE LOST THIS TIME, BUT NOT BY MUCH. DON'T TAKE IT TOO HARD.
YOU'RE RIGHT.
YOU'RE ABSO-LUTELY RIGHT.
WHO KNEW THAT A SINGLE PIECE OF PAPER...
COULD CUT AS DEEP AS A KNIFE?
THANK YOU SO MUCH, SIR.

HEY, SARASA!
I FORGOT TO ASK BEFORE!
IS YOUR BOY-FRIEND GONNA COME WATCH THE SHOW?
THE WATER WAS KINDA HOT TODAY, EH?
HUH?! THAT'S A GOOD IDEA! I'D LOVE FOR HIM TO COME, BUT--
HE'S NOT COMING.
CRAP, I FORGOT MY MOISTURIZER.
YOU CAN USE MINE!
WHY'D YOU ANSWER FOR HER, NARACCHI?
I DUNNO.
I DON'T REALLY CARE ABOUT HIM...
BUT I BET...
HE'S NOT COMING.
I DON'T THINK THIS GUY IS EVEN REAL!
RIGHT, SAWA?!
......
OH, JEEZ.

THIS HAS REALLY GOTTEN UNDER MY SKIN.
MORNING.
KOUKA
OH!
TAKEI-SENPAI! GOOD MORNING!

LOOKING AT THESE?
A FAN STOPPED ME ON MY WAY HERE AND HANDED THEM TO ME.
I'M NOT SURPRISED YOU HAVE FANS ALREADY.
TALK ABOUT GETTING ON THE BANDWAGON EARLY!
IT'S JUST BECAUSE I'M MY CLASS REP.
AND I WAS AT THE SPORTS FESTIVAL, TOO.
I'M SORRY YOU DIDN'T GET THE ROLE.
BEING A MODEL STUDENT IS SO BORING, HUH?
EVERYONE EXPECTS YOU TO BE PERFECT...
AND IF YOU MESS UP EVEN ONCE, THEY'LL NEVER FORGET IT.
EVEN IF THEY'LL REMEMBER WHEN THE SCREW-UP DOES SOMETHING RIGHT. IT'S SO UNFAIR.

AND WHEN YOU START REPPING THE SECOND-YEARS, YOU'LL BE REPRESENTING THE WHOLE SCHOOL.
YOU'LL ALWAYS HAVE TO BE ON YOUR BEST BEHAVIOR.
YOU CAN'T HELP BUT FEEL JEALOUS OF PRODIGIES OR STRONG PERSON-ALITIES.
YOU FEEL THAT WAY TOO SOMETIMES, TAKEI-SENPAI?
ALL THE TIME!
I MEAN, LOOK AT MY ASSISTANT REP!
THAT GIRL HAS A WAY BIGGER PERSON-ALITY THAN I DO!
HA HA!
BUT SHE SAYS THE THINGS I CAN'T, WHICH IS A HUGE HELP TO ME.

TO FANS OF THE TROUPE...
WE'LL *ALWAYS* BE THE CLASS REPS OF OUR RESPECTIVE CLASSES.
IT'S A LOT OF PRESSURE.
BUT THE SKILL YOU'VE WORKED SO HARD TO CULTIVATE...
WILL *NEVER* FAIL YOU.
SURE, IT'S NERVE-RACKING,
BUT NOW...
I'M *FINALLY* GOING TO BE ABLE TO STAND ON THE STAGE I'VE DREAMED OF THIS WHOLE TIME.
I'M SO EXCITED.
THAT'S WHAT THESE TWO YEARS HAVE ALL BEEN ABOUT.

I'M SURE YOU'LL FEEL THE SAME WAY IN A YEAR'S TIME.
I'M GLAD YOU'RE THE CLASS REP FOR THE FIRST-YEARS.
YOU'VE KEPT YOUR CLASS ORGANIZED, AND THAT MADE THINGS ALL THAT MUCH EASIER ON ME.
THANK YOU.

PLIP
HUH?
OH, JEEZ!
PLIP PLIP
AH HA HA HA!
DON'T CRY NOW!
SAVE IT FOR GRADU-ATION!
HUH?!
UGH, I HATE CRYING IN FRONT OF OTHER PEOPLE!
I'M SO EMBAR-RASSED.
I FEEL YOU. IT'S SO HUMILIATING TO CRY IN FRONT OF OTHER PEOPLE.
WE'RE PRETTY SIMILAR, HUH?

THEATRICAL
ART
TODAY...
I'LL BE THE SAME "GOOD GIRL" I WAS BEFORE...
AND REPRESENT THE HUNDREDTH CLASS OF THE KOUKA SCHOOL OF MUSICAL AND THEATRICAL ARTS.
JUST LIKE EVERY OTHER DAY.

SAWA!
ARE YOU DONE CLEANING?
I JUST FINISHED.
ME TOO!
HEY, SAWA?
THANK YOU FOR GIVING ME THAT LITTLE PUSH DURING AUDITIONS.
IF YOU HADN'T, I *NEVER* WOULD HAVE GOTTEN THE ROLE.
SO, UM...
I'VE NOTICED YOU'VE BEEN KIND OF DOWN LATELY.

I KNOW I MIGHT NOT BE ABLE TO DO MUCH...
BUT PLEASE LET ME KNOW IF I CAN HELP!
Really, if I can do anything ...!
JOLT
OH, BOY!
C'MON, NOW, DON'T CRY!
I'M SORRY! I JUST GOT ALL WORKED UP...
HEE HEE!
THANKS.
PERHAPS...

IF SALIERI HAD HAD A FRIEND TO SUPPORT HIM...
HE WOULD HAVE NEVER GIVEN IN TO BITTERNESS AND DESPAIR.
YOU WANT TO KNOW WHY I CHOSE WATANABE-SAN FOR TYBALT?
ANDOU-SENSEI TOLD YOU I MADE THAT CALL?
YES, MA'AM. FORGIVE ME FOR ASKING SO BLUNTLY.
BUT I WAS HOPING TO LEARN SOMETHING FROM IT.
I SEE. VERY WELL, THEN.
HERE'S THE REASON I CHOSE WATANABE-SAN OVER YOU.
GULP

BECAUSE HER AUDIENCE APPEAL LEVEL...
WAS JUST A *BIT* HIGHER THAN YOURS!
?
IT WAS DIFFICULT TO DISCERN WHICH OF THE TWO WOULD RATE HIGHER...
BUT CONSIDER THIS-- EIGHTY PERCENT OF KOUKA'S AUDIENCE ARE WOMEN.
WE MUST CONSIDER WHAT *THEY* WANT OUT OF TYBALT.

YOUR TYBALT ...
WHO DESTROYS HIMSELF THROUGH HIS LOVE, WAS WONDERFUL INDEED.
HE LOVES JULIET (ME) SO MUCH HE'S GONE TO THE DARK SIDE!
BA-DUMP
BUT WATANABE-SAN'S TYBALT, WHO OPENED UP ABOUT HIS FEELINGS AT THE VERY END...
WOULD RESONATE MUCH MORE STRONGLY WITH OUR AUDIENCE.
HE ONLY SHOWS HIS TRUE FEELINGS TO JULIET (ME)!
BA-DUMP
I SUPPOSE TO PUT IT CONCISELY ...
WATANABE'S TYBALT HAD *MUCH* MORE FANGIRL APPEAL!
FANGIRL APPEAL ?!!

SWAY
FAN-GIRL.
FAN-GIRLING.
THAT ONE WORD MADE EVERYTHING COME TOGETHER.
UH?
I'M SORRY, ANDOU-SENSEI. YOU TRIED REALLY HARD TO EXPLAIN IT.
BUT IT MAKES SO MUCH MORE SENSE...
WHEN YOU EXPLAIN IT BY FANGIRL POTENTIAL ALONE!
OHGI-SENSEI IS TRULY A DIEHARD KOUKA FAN!
SUGI-MOTO-SAN!

ARE YOU HEADING HOME?
WE ARE, TOO!
I CAN'T BELIEVE IT'S ALMOST NEW YEAR'S ALREADY!
YOU'RE RIGHT. IN NO TIME AT ALL, WE'LL BE MOVING TO THE SECOND-YEARS LOCKER ROOM.
I HOPE I CAN BE A GOOD BIG SISTER!
OOH!
LOOKIE OVER THERE!!
THAT'S THE POSTER THEY PHOTOGRAPHED US FOR!

OH, WOW! YOU LOOK SO COOL RIGHT IN FRONT THERE, SUGI-MOTO-SAN!
COME TO THE STAGE OF YOUR DREAMS!
KOUKA SCHOOL OF MUSICAL AND THEATRICAL ARTS
APPLY TODAY
YEAH, YOU LOOK COOL.
THEY MADE ME CROUCH DOWN TO FIT IN BETTER.
OH, RIGHT, I FORGOT!

SARASA.
CONGRATS ON GETTING THE ROLE!
SUGIMOTO-SAN!
WOW, SO THEY WENT WITH PINK FOR THE RECRUITING POSTER THIS YEAR?

LOOK HOW TINY AND CUTE YOU GUYS ARE. YOUR SKIN LOOKS GREAT, TOO.
I'M SO JEALOUS.
?! ?! ?!
MITSUKI KEITO-SAMA, TOP STAR OF THE AUTUMN TROUPE, AND RYUUZAKI MARIA-SAMA?!
HUH?
WAIT.
ARE YOU...
UM, HI...?
THE GIRL STANDING FRONT AND CENTER THERE?
Y-YES, MA'AM!
SO YOU'RE THE FIRST-YEAR CLASS REP, THEN.
YES, MA'AM!

YOU'RE SO LOVELY AND GRACEFUL.
YOU COULD BE AN AMAZING OTOKOYAKU OR MUSUMEYAKU.
BREAK A LEG, HON!
I AM SO FREAKING GLAD I'M THE CLASS REP!!
SUGIMOTO-SAN, YOUR NOSE IS BLEEDING.

22

KAGEKI SHOJO!!

ON JANUARY 1ST...
AT THE KOUKA THEATER TROUPE'S MAIN THEATER IN KOBE...
KOUKA
Kouka Theater

MEMBERS OF THE TROUPE, STAFF, AND OTHER ASSOCIATES HOLD A CEREMONY TO RING IN THE NEW YEAR.
KOUKA
New Year's Kagami Mochi Ceremony
THEN THEY HOLD THE NEW YEAR'S KAGAMI MOCHI CEREMONY, WHERE FANS ARE CALLED UP TO PARTICIPATE.
HAKUSEN
HAKUSEN

RIGHT AFTER THAT IS OPENING NIGHT FOR THAT YEAR'S SHOW.
BUT FOR US STUDENTS...
SO NEW YEAR'S DAY IS AN INCREDIBLY BUSY DAY FOR THE TROUPE.

ONE DAY EARLIER, ON DECEMBER 31ST...
WA HA HA HA!
WE'RE ENJOYING OUR WINTER BREAK, WITH MOST OF US RETURNING HOME FOR NEW YEAR'S.
I'M SO SORRY, SARASA!
WE WON'T BE ABLE TO GO TO THE CULTURE FESTIVAL ANYMORE!
AWW! WHY NOT?
WELL, YOU SEE, WE JUST FOUND OUT WE'RE PREGNANT AGAIN...
I'M GONNA BLUSH!
WHAT?! OH MY GOSH, CONGRATS !!!

YAKKUN, YOU'RE GONNA BE A BIG BROTHER!
BIG BUDDER?
OUR DOCTOR TOLD US NOT TO GO ON ANY BIG TRIPS UNTIL SHE'S INTO HER NEXT TRIMESTER.
STILL, WE'RE SORRY.
AWW, IT'S OKAY!
BUT THEN WHO'S GONNA GO WITH GRANDPA KEN TO THE FESTIVAL?!
SNRRRK
SNRRK
SNRK
IT'S HIS FIRST TIME IN KOBE, RIGHT?
OH.
I COULD GO WITH HIM.

WAIT, REALLY?! YOU SURE, AKIYA?!
YUP.
I HAVE SOME TIME OFF AROUND THEN.
THEN THAT'S GREAT! THANKS A BUNCH!
I'M WORRIED.
SHIRAKAWA-SAN IS SOMEWHAT FAMOUS.
SO I WORRY.
SHIRAKAWA... SAN?
AW, THANKS, BUT I'LL BE OKAY.
GRIN
I'M NOT WORRIED ABOUT YOU, SHIRAKAWA-SAN.
I'M WORRIED ABOUT ANY WEIRD RUMORS GETTING STARTED ABOUT SARASA.
REALLY WORRIED.
......
HEY!
AI-CHAN, LOOK!

*Kohaku is a yearly TV song competition in Japan and is a New Year's tradition.

WHAT DID YOU THINK OF JPX48'S PERFORMANCE, KOUZABUROU-SAN?
EVERY SINGLE ONE OF THEM IS OVERFLOWING WITH YOUTH AND LIVELY PEP.
AND WHEN THEY COME TOGETHER WITH THOSE NUMBERS, IT'S ALMOST OVERWHELMING!
ABSOLUTELY PHENOMENAL TO EXPERIENCE LIVE!
KOUZABUROU...
OH YEAH, HE DID SAY HE WAS GONNA BE ON THAT THIS YEAR!
ZZZ...

OH!
FALLING FROM THE SKY LIKE WHIIIIITE SNOW... ♪
BUT WARM... ♪
LIKE A SPRING FAIRY! ♪

AH, THE KOUKA CHERRY BLOSSOMS!
GYA HA HA!
THAT'S IT!
THEY'RE TOTAL PROS!
WITH THAT SAID AND DONE...
FALLING UPON THE EARTH, BECOMING ONE WITH THE GROUND...
TO BECOME NEW LIFE...
AND GO TO YOU.
NOW THEY FLUTTER AND DAAAANCE!

WE GREETED THE NEW YEAR.
HAPPY NEW YEAR TO YOU ALL.
LET'S MAKE THIS ANOTHER WONDERFUL YEAR, TOGETHER.
THE SECOND-YEARS FOCUSED THEIR EFFORTS ON THE CULTURE FESTIVAL AND GRADUATION.
NOW, GIRLS!
DON'T LIFT THAT TOE!
SOFTEN THAT FOOT! YOU'RE NOT GRADUATING WITH TAPS LIKE THAT!
SMILE!
LOOK OVER THERE.
TIME SPED BY AT A DIZZYING PACE.
TAP, TA-TAP!
TA-TA-TA-TAP!
TAP!
TA-TAP!
TA-TAP!
THOSE ARE THE PRO-DUCERS FOR SPRING AND WINTER!
THEY'RE HERE TO OBSERVE!

I WONDER WHICH TROUPE I'LL END UP IN!
SPRING'S FOR JAPANESE DANCE, SUMMER'S FOR MODERN DANCE, AUTUMN'S FOR SINGERS, AND WINTER'S FOR ACTING.
I HEARD SUMMER'S THE ONLY TROUPE LOW ON MUSUMEYAKU, YEAH?
WINTER.
I HOPE I GET PICKED FOR WINTER.
SORRY, EVERYONE, I HAVE TO RUN! GOOD WORK TODAY!
GRIN
OH HO!
HIJIRI...
SURE STANDS OUT.

SHE'LL DEFINITELY BECOME A TOP STAR.
BE-BEEP
Sorry, Hijiri
I can't come to the culture festival
I'm really sorry
MY BELOVED ROMEO HAS KILLED MY DEAR COUSIN? NAY!
WHAT CRUEL TRICK OF FATE!
FALSE-HOODS!
EVEN SO, I LOVE YOU TRULY...
JULIET!

STOP STEPPING FORWARD EVERY TIME ONE OF YOU IS SUPPOSED TO SPEAK.
IT'S AMATEUR.
YOU'RE NOT IN ELEMENTARY SCHOOL ANYMORE.
......
OKAY.
THEN COULD WE HAVE A SPOTLIGHT SHINE ON EACH OF US AS WE SAY OUR LINES?!
ONLY THE STARS GET TO HAVE SPOTLIGHTS ON THEM!
SHEESH, WATANABE, YOU SAY THE DARNDEST THINGS.
SO... ANDOU-SENSEI?
ARE WE PERFORMING BETTER NOW THAN WE DID AT OUR AUDITION?
YEAH, I'D SAY YOU'VE IMPROVED.

STILL...
SKITS LIKE THIS ARE BARELY MORE THAN A SCENE.
AND YOU DON'T HAVE LOTS OF LINES EITHER, YEAH?
IT'S EASY TO OVERTHINK THINGS, GET TOO DEEP INTO YOUR HEAD, AND RUIN EVERYTHING.
THE ONLY THING YOU SHOULD BE WORRIED ABOUT RIGHT NOW IS THE PACING BETWEEN YOUR SETS OF LINES.
WHAT ELSE...? *OH*, ONE MORE.
THEY SAY THAT *PRESENCE* IS SOMETHING YOU CONSCIOUSLY CREATE...
BUT WITH YOUR LACK OF EXPERIENCE, THE BEST PRESENCE YOU CAN PLACE ON STAGE IS YOUR *YOUTH*.

AND THAT WAS IT FOR HIS CRITIQUE!
SO NOW WE'RE GONNA MEET ONCE A WEEK TO REHEARSE WITH THE REST OF THE FESTIVAL PROGRAM.
COOL.
AND ANDOU-SENSEI IS TOTALLY RIGHT.
EVER SINCE I GOT IN, I'VE BEEN THINKING ABOUT TYBALT WAY TOO HARD!
HOW HE LOVES JULIET ...
BUT MAYBE HIS RAGE TURNS INTO LOVE FOR ROMEO, AND NOW I'M ALL CONFUSED...
LIKE SOME CHEAP ROMANCE NOVEL...
?
WOW...
IT'S JUST LIKE AT THE SPORTS FESTIVAL.
"DON'T THINK TOO HARD AND MESS UP."

"PRESENCE IS SOME-THING YOU CONSCIOUSLY CREATE."
HUNH.
ALL WE HAVE TO DO IS PERFORM OUR ROLES AND SUPPORT THE SECOND-YEARS.
IF YOU'RE NERVOUS, I'LL WATCH YOU PRACTICE.
YOU'RE AN ANGEL, AI-CHAN!
IT'S NO BIGGIE!
WE'RE ROOM-MATES.
AND FRIENDS.
ROLL

LET'S ALL GIVE IT OUR BEST AND SEND THE SECOND-YEARS OFF IN STYLE!

I CAN'T BELIEVE THE FESTIVAL'S TOMORROW!

GRANDPA KEN'S GOT A DELIVERY TO MAKE IN THE MORNING...
SO WE'LL MISS THE START.
BUT WE'LL *DEFINITELY* BE THERE FOR YOUR PERFORMANCE.
YAY!
I'M GONNA PUT ON A PERFORMANCE THAT'LL SHOW YOU I'M WORTHY!
WORTHY?
SHOW HIM SHE'S WHAT?

GOOD MORNING!
Ensemble Theater
WELCOME TO THE KOUKA SCHOOL OF MUSICAL AND THEATRICAL ARTS CULTURE FESTIVAL!
Kouka Theater
KOUKA
99th Class of the Kouka School of Musical and Theatrical Arts Graduation CULTURE FESTIVAL

ENTER HERE FOR THE CULTURE FESTIVAL PERFORMANCE!
WE'RE EXCITED TO PERFORM FOR YOU!
OH, WOW!
THEY'RE SO CUTE! ♡
YOUR CULTURE FESTIVAL IS TODAY?
WE'RE HERE TO SEE THE SUMMER MATINEE AND DON'T HAVE TICKETS FOR THE CULTURE FESTIVAL...
BUT BREAK A LEG, DEARS! YOU'RE ALL FUTURE STARS!
THANK YOU VERY MUCH!
COULD WE TAKE A PICTURE OF YOU?
OF COURSE!

YOU CAN PURCHASE PAMPHLETS FOR THE CULTURE FESTIVAL HERE!
THANK YOU FOR COMING TODAY!
Ensemble Theater
Is my smile driving people away?
It's fine! I saw you practicing earlier, you're doing great!
LOOK AT THE SAWADA TWINS, STEALING THE SPOTLIGHT.
THEY'RE SO CUTE!
TWINS!
TWINS!
MAY I HAVE ONE, PLEASE?
OH!
YES, OF COURSE!

YOU'RE FIRST-YEARS NOW, AREN'T YOU?
HOW TALENTED YOU MUST BE TO GET IN--AND FOR THE HISTORICAL 100TH CLASS, TOO!
I'LL BE HERE FOR YOUR CULTURE FESTIVAL NEXT YEAR.
BREAK A LEG, OKAY?
WE WILL!
THANK YOU VERY MUCH!
UMM, COULD I LIKE, GET A PAMPHLET?
OF COURSE! THANK YOU SO MUCH!!
......
UH...?

UGH, I WANTED TO GET ONE FROM NARACCHI.
IS SHE, LIKE, HERE?
NARATA-SAN?
COULD I BORROW YOU FOR A MOMENT?
BACK-STAGE.
YES?
SHOOP
THIS FRILL ON MY COSTUME HAS STARTED TO COME LOOSE.
I'LL TAKE CARE OF IT.
WHIP
I'VE BECOME A SEWING MASTER IN THIS PAST YEAR.
IT'S YOUR BIG DAY.

I'M GOING TO DO WHAT-EVER I CAN...
TO MAKE YOUR CULTURE FESTIVAL A GOOD ONE.
RIGHT. THANKS.
FLOAT,
......
I'LL WORK HARD TO HELP YOU, TOO!!
COULD YOU GET THAT FOR ME?
YUP!

DIM
LADIES AND GENTLEMEN.
WELCOME TO THE KOUKA SCHOOL OF MUSICAL AND THEATRICAL ARTS 99TH ANNUAL...
CULTURE FESTIVAL PERFORMANCE.

THIS SHOW IS THE RESULT OF TWO YEARS OF HARD WORK, FRIENDSHIP, AND MOST OF ALL, LOVE FOR OUR CRAFT.
TODAY, WE'LL SHOW OUR GUESTS AND ESTEEMED FAMILY MEMBERS...
JUST HOW MUCH WE'VE GROWN IN OUR TIME HERE!
YAAY!

WAIT HERE. I'LL BE BACK IN A FEW MINUTES.

GRAND-PA KEN!
IT'S ME, AKIYA!
DING-DONG
ピンポーン
HELLO?

GRANDPA?

RISA-SENPAI! YOU WERE AMAZING OUT THERE!
YOUR CARMEN WAS SO PASSION-ATE AND SEXY!
YOU THINK SO?
YOU WERE PERFECT FOR THE ROLE!
MM-HMM.
YOU'VE GOT A HOT BOD, RISA. IT'S THE PERFECT ROLE FOR YOU.

YOU'VE GOT *SUCH* A POWERFUL STAGE PRESENCE!
YOU SHOULDN'T EVEN *TRY* TO BE A SWEET, INNOCENT MUSUME-YAKU!
YOU'D BE *SO* MUCH BETTER AS A DARK, HORRIBLE VILLAINESS!
WOW, HIJIRI.
I WAS *JUST* THINKING THE SAME THING!
UHHHHHH!
I WAS *JUST* THINKING HOW PERFECT YOU ARE FOR THE UNSCRUPULOUS, SELFISH, DOMINEERING SCARLETT!
SARASA.

THE FIRST-YEARS ARE UP AFTER THIS NEXT PERFORMANCE. YOU NEED TO GO LINE UP.
OH!
OKAY!!
BREAK A LEG OUT THERE.
WILL DO!
......
BREAK A--
WATA-NABE-SAN!
WATANABE SARASA-SAN!
YES, MA'AM!

SOMEONE FROM HOME JUST CALLED.
YOUR GRANDFATHER WAS FOUND UNCONSCIOUS.

HUH...?
WHAT?
THAT'S ALL THEY SAID.
WHAT WOULD YOU LIKE TO DO? WOULD YOU LIKE TO GO BE WITH YOUR FAMILY?
BUT...
BUT I'M GOING TO PERFORM.
I HAVE LINES.
AND THEY DEPEND ON ME...
YOU CAN GO.

NOBODY'S HERE TO SEE YOU OR EVEN HOPING TO SEE SOMETHING FROM YOU.
NO ONE'S MISSING OUT.
ANYONE CAN STAND IN FOR YOU.
SO GO.

23

JUST GO. NOBODY CAME HERE TO SEE *YOU* PERFORM.
NO ONE'S MISSING OUT.
WHAT THE HELL, HIJIRI?! HOW CAN YOU SAY THAT?!
BUT...
BUT I HAVE TO GET ON STAGE...
THEY PICKED ME.
I HAVE A RESPON-SIBILITY--
TO WHAT?

YOU'VE ONLY GOT ONE RESPON-SIBILITY.
AND IT'S NOT EVEN TO GET UP ON STAGE.
IT'S TO SAY YOUR LINES WITHOUT SCREWING UP.
THAT'S WHY I'M TELLING YOU...
TO GO HOME, RIGHT NOW.
......
I'M SORRY.

I WANT TO BE WITH MY FAMILY.
I'M SO SORRY.
PLEASE LET ME LEAVE.

HIJIRI!!
THAT WAS COMPLETELY OUT OF LINE, AND YOU KNOW IT!
AND YOU DIDN'T HAVE TO SAY IT LIKE THAT, EITHER!
COULD YOU *NOT* DO THIS RIGHT BEFORE I'M ABOUT TO PERFORM?
A COLD-HEARTED GIRL LIKE YOU DOESN'T GET STAGE FRIGHT!
YOU'RE RIGHT. THAT'S PART OF WHAT MAKES STARS SO GREAT.
SORRY. I JUST CAN'T BRING MYSELF TO LIE.

HIJIRI...
YOU'RE SUCH A--
IT'S MY TIME TO SHINE!!
WHY WOULD I BE NERVOUS?!
THAT'S SUCH A RIDICULOUS THOUGHT!
OOH!
IT'S ALMOST TIME!
I'M SO EXCITED ...
I'M SO, SO EXCITED!

GOSH, IF ONLY... THIS CULTURE FESTIVAL COULD LAST FOREVER!
WATCH ME, RISA.
RIGHT NOW, THAT STAGE...
IS ALL MINE!!

WATANABE DID WHAT?!
IS THAT RIGHT? OKAY.
SARASA'S NOT GONNA PERFORM?
OH NO!
MMKAY, WHO WANTS TO STAND IN FOR HER?
Quick Shift
HUH?!
WELL, IT IS THE ONLY WAY FORWARD.
SUGIMOTO, CAN YOU HANDLE IT?
HUH?
S-SURE!
BUT WE'VE NEVER PRACTICED WITH SAWA.
DO YOU THINK IT'LL WORK?

ANDOU-SENSEI.
HUH?
WHAT IS IT, NARATA?
I'LL...

I'LL...
PLAY TYBALT.

WHOA...
WHOA.
WHOA!!

WHAT GIVES *YOU* THE RIGHT TO COME OUT HERE AND SAY YOU'LL DO IT, NARACCHI?
'CAUSE YOU'VE GOT EXPERIENCE BEING ON A BIG STAGE ALREADY OR SOMETHING?
YEP.
OKAY, SO WHAT?!
SAWA'S JUST AS GOOD!
SHE'S PERFORMED IN FRONT OF PEOPLE DOZENS OF TIMES AT BALLET COMPETITIONS!
AND *SHE* ACTUALLY WANTED TO PLAY TYBALT IN THE FIRST PLACE!

I'VE BEEN HELPING SARASA PRACTICE FOR WEEKS.
I KNOW EVERYTHING.
SARASA'S PERFORMANCE, BREATHING, TIMING...
I CAN DO IT ALL PERFECTLY.
THAT, AND...
IF...
IF I SCREW UP SOMEHOW...

Naracchi same as ever
komomo is CUTE
would it literally kill this girl to smile???
why's naracchi in the front?
love the new song so much
new son
KOMOMO IS MY WAIFU
I'm so bored of JPX now tbh
← lol okay bye
The era of MiriMiri starts NOW!!
I HAVE WAY MORE EXPERIENCE MAKING A FOOL OF MYSELF IN FRONT OF AN AUDIENCE THAN ANY OF YOU.
PLUS...
Naracchi: "Not smiling for you, nope!!"
I'M ALSO USED TO BEING HATED AND FLAMED.
SO IF THIS GOES WRONG, IT CAN ALL BE PINNED ON ME.

THAT'S WHY I SHOULD DO THIS!
HONK
THE PHONE NUMBER YOU HAVE DIALED...
IS OFF OR OUT OF SERVICE.

History
Akiya-kun
Akiya-kun
Akiya-kun
Akiya-kun
Akiya-kun
Akiya-kun
Pwip
Pwip
Sarasa-chan
I just saw you called a bunch
Sorry I couldn't pick up
Pwip

Sorry I couldn't pick up
Shou-chan fre out and callec the school
JOLT
"GRANDPA KEN'S GONNA SPEND THE NIGHT IN THE HOSPITAL.

"BUT HE'S OKAY. APPARENTLY HE JUST OVEREXERTED HIMSELF. HE'S GOING TO BE OKAY.
"I'M SORRY WE COULDN'T COME SEE YOU PERFORM."

I'm
come see
Break a leg on
stage!!!!

LADIES AND GENTLE-MEN...
THIS NEXT PERFORMANCE WILL CAP OFF OUR FIRST HALF BEFORE INTERMISSION.
PLEASE ENJOY THIS SPECIAL PERFOR-MANCE BY THE HUNDREDTH CLASS.
A STUNNING CHORAL ARRANGE-MENT...
AND A SHORT READING FROM *ROMEO AND JULIET*.

A SHORT READING BY THE FIRST-YEARS?
THEY'VE NEVER DONE THAT BEFORE.
MAYBE THEY'RE CHANGING THINGS UP FROM HERE ON OUT.
OH!
LOOK!
THEIR NAMES ARE HERE IN THE PAMPHLET!
OOH!
THE GIRL WHO'S PLAYING TYBALT, WATANABE-SAN...
SHE WAS THE GIRL WHO HELD HANDS WITH SEI-SAMA AT THE SPORTS FESTIVAL!
First-Year Chorus & Reading
Romeo and Juliet
SONG: Beyond the Reach of Love
Romeo: Shirota Minami
Juliet: Yamada Ayako
Tybalt: Watanabe Sarasa
Nurse: Hamada Asami

OHHHH!
I LOVED THAT MOMENT FROM THE SPORTS FESTIVAL! I WATCH IT ON THE DVD ALL THE TIME!
THAT'S THE REALLY TALL GIRL, RIGHT?
I BET SHE HAS AMAZING GRADES.
I WONDER IF SHE'S A GOOD ACTRESS.
HEY!
THERE YOU ARE!
HOW'RE THE FIRST-YEARS? NERVOUS?
I GUESS.
FIGURES.
THIS IS THEIR FIRST TIME PERFORMING FOR AN AUDIENCE!

NOW, WITHOUT FURTHER ADO...
IT'S THE FIRST TIME I'M SEEING MY DARLING NIECE PERFORM ON THE KOUKA STAGE, TOO.
I'M FEELING NERVOUS FOR HER!
I'M EVEN WEARING A SUIT!
BUT I KNOW THAT EVEN IN THAT HUGE CHORUS...
MY PRECIOUS AI IS GOING TO SHINE LIKE A DIAMOND!
THE 100TH CLASS OF THE KOUKA SCHOOL OF MUSICAL AND THEATRICAL ARTS...
HA HA HA!
I'M SURE SHE WILL.
WILL NOW PERFORM FOR THE VERY FIRST TIME!

......
HUH?!
MURMUR
HUH...?
NO WAY!
THIS WASN'T IN THE PAMPHLET...
ISN'T THAT...

NARACCHI?

Side Story:
Nakayama
Risa

LOOK! IT'S THE NEW TOP MUSUME-YAKU OF THE WINTER TROUPE...
NOJIMA HIJIRI-SAN!!
SHE'S SO LOVELY. NO WONDER SHE'S A TOP STAR AFTER JUST ONE YEAR OF PERFORMING!
SHE MAKES A PERFECT PAIR WITH SEI-SAMA...
BEAUTIFUL AND TALENTED!

OOF, THAT TREE IS IN THE WAY.
I CAN'T SEE SEI-SAMA AND HIJIRI-SAN!
WHY IS THERE EVEN A TREE THERE?!
GET OUT OF THE WAY!!
GASP
......
WHAT A NIGHT-MARE!
AFTER MAKING IT THROUGH THE CULTURE FESTIVAL, US SECOND-YEARS FINALLY GET THE CHANCE...

TO ENJOY A SHORT BREAK...
BEFORE THE TENSION AND WORRY OF GRADUATION AND JOINING THE TROUPE.
THE CULTURE FESTIVAL WAS SO MUCH FUN!
KNOWING WE MIGHT SOON BE SEPARATED INTO DIFFERENT TROUPES...
IT WAS SO PERFECT!
THAT WAS THE FIRST AND LAST TIME WE'LL ALL GET TO PERFORM TOGETHER AS A CLASS.
WE ALL STRUGGLED TO COME TO TERMS WITH OUR EMOTIONS.
BOY, THOUGH, THAT FIRST-YEAR PERFORMANCE...
YEAH. THE AUDIENCE WAS ALL ABUZZ ABOUT NARACCHI.
IT WASN'T DONE ON PURPOSE, YOU KNOW.

WE NEEDED A STAND-IN, AND NARATA-SAN HAPPENED TO BE THERE.
BUT EVEN IF IT WAS JUST CHANCE...
YOU CAN'T DENY THAT NARATA-SAN...
HAS ALWAYS HAD WHAT IT TAKES.
TEE HEE!
CRACKLE

OH, HEY!
HAVE YOU GUYS PICKED OUT YOUR STAGE NAMES YET?
I DID!
I HAVEN'T YET!
HEY TAKEI, WHAT'D YOU PICK?
ASAGIRI RYOU.
I'M GOING WITH NAGI JULIA.
I CAN'T CHOOSE BETWEEN MIKU AND SHIZUKU.
HOW ABOUT YOU, RISA?
WELL...
I'M KINDA EMBAR-RASSED TO TELL YOU GUYS.
AHA HA!
WHAT'RE YOU TALKING ABOUT, RISA?
YOU *NEVER* GET EMBAR-RASSED!
GRR!

NO NEED TO PRETEND TO BE BASHFUL!
FINE, THEN, WHAT'S YOUR STAGE NAME GOING TO BE, HIJIRI?
HEE HEE...
I'M NOT TELLING. ♡
WHAT ?!
IT'S NOT LIKE I'M EMBARRASSED!
I JUST DON'T WANT TO REVEAL IT YET.
YOU'LL FIND OUT SOON ENOUGH!
ALL RIGHT, ENOUGH.
TIME TO HEAD TO THE NEXT CLASSROOM!
GOD, YOU REALLY PISS ME OFF!
AWW, COME ON, NOW! LOVELY KOUKA FAIRIES SHOULDN'T USE SUCH FOUL LANGUAGE!
THAT MEANS YOU TWO, TOO!

WHERE DOES SHE GET OFF, ACTING LIKE SHE'S ALL THAT?!
WHO?
OH, YOU MEAN HIJIRI?
SHOOT, I SAID THAT OUT LOUD!
HIJIRI'S GRADES HAVE BEEN GOOD THIS WHOLE TIME.
AND, WELL, SHE'S GOT THAT DISNEY PRINCESS FACE.
SHE MIGHT BE A BITCH, BUT I CAN SEE WHY SHE'S SO COCKY.
DO YOU WANT TO BE A TOP STAR, HINANO?
ME?
I'M NOT MUCH FOR PLAYING HEROINES.
FOR ME, I JUST WANT TO PERFORM IN STAGE MUSICALS.
AND I LOVE KOUKA, OF COURSE.
AND AS LONG AS YOU GET INTO KOUKA, YOU CAN GET ON STAGE RELATIVELY FAST.
AND YOU GET TO TAKE LESSONS FROM SOME OF THE BEST TEACHERS IN THE BUSINESS!

THEY SAY IT TAKES TEN YEARS FOR OTOKOYAKU TO REACH THE HEIGHTS OF THEIR ABILITY...
BUT MUSUMEYAKU PEAK WITHIN THEIR FIRST SIX YEARS.
SO I'LL PROBABLY QUIT KOUKA SOMETIME IN MY TWENTIES.
THEN I'LL TRY OUT FOR A DIFFERENT THEATER TROUPE.
WOW.
I NEVER CONSIDERED THAT PATH BEFORE.
DO YOU WANT TO BE A TOP MUSUMEYAKU, RISA?
I DON'T KNOW.
I JUST...

I WANT TO CROSS THAT SILVER BRIDGE MORE THAN JUST THE ONE TIME AT OUR DEBUT PERFORMANCE, YOU KNOW?
IN MARCH, WE'LL GRADUATE FROM THE KOUKA SCHOOL OF MUSICAL AND THEATRICAL ARTS...
THEN AUTO-MATICALLY JOIN THE KOUKA THEATER TROUPE.
AND IN APRIL, WE'LL DEBUT.

AT THAT DEBUT PERFORMANCE, ALL THE NEW MEMBERS WILL PERFORM A LINE DANCE...
ON THE SPECIAL SPACE IN FRONT OF THE STAGE: THE SILVER BRIDGE.
AND FOR MANY OF THE GIRLS IN THAT LINE...

THAT WILL BE THE FIRST AND LAST TIME THEY EVER WALK ACROSS THAT SILVER BRIDGE.
I KNOW THAT HIJIRI...
nom nom

WILL CROSS THAT BRIDGE MORE THAN ONCE.
I HATE TO ADMIT IT...
BUT WHEN SHE PERFORMED AT THE CULTURE FESTIVAL...
SHE WAS TERRIFYINGLY BEAUTIFUL.

"WOW, RISA, YOU'RE GOING TO TRY OUT FOR KOUKA?"
"YEAH, I WANT TO BE A MUSUME-YAKU!"
"I'VE ACTUALLY TRIED OUT A FEW TIMES BUT DIDN'T GET IN."
"YOU'RE SO PRETTY, RISA! I KNOW YOU'LL MAKE IT SOMEDAY!"
"I'LL KEEP MY FINGERS CROSSED!"
"THANKS! I'M GONNA KEEP AT IT!
"BUT I CAN ONLY APPLY TWO MORE TIMES BEFORE I'M TOO OLD."
Application Status
NOT ADMITTED
Thank you for applying.
"MAN THOUGH, NAKAYAMA WOULD LOOK WEIRD IN A FRILLY DRESS!"
"WHEN I THINK OF HER DANCING, I THINK TOPLESS, SHAKING HER BOOTY."
"WHAT THE HELL?!"
"GO TO HELL, JERKS!"

"MAYBE I'M NOT CUT OUT FOR IT..."
"HEY."
"DON'T YOU GO SAYING YOU'RE QUITTING NOW, ALL RIGHT?
"YOU'VE GOT ONE MORE CHANCE.
"YOU WANT TO BE A CUTE HEROINE, YEAH?
"BESIDES! YOU'RE A CUTIE, JUST LIKE YOUR MA."
"I KNOW YOU'LL GET IN NEXT TIME!
"NO.
"I KNOW I LOOK MORE 'MATURE' THAN 'CUTE'..."
"HEY!
"LET'S THINK OF A FUN STAGE NAME FOR YOU!

"YOU WANT TO BE A TOP STAR, RIGHT?
"THEN WHY DON'T YOU PICK A STAR FOR YOUR NAME?
"THE TOP STAR IS THE STAR THAT SHINES BRIGHTEST ON HER OWN!
"WHICH STAR WOULD BE PERFECT FOR YOU?
"OH, HOW ABOUT ONE FROM THE BIG DIPPER?"
GH...
SNIFF ...
GAAAAH!

I'M... I'M...
I CAN'T BELIEVE THIS IS MY LAST DAY WITH YOU AS MY BIG SISTER...
I'M SOOOO SAD!
UH...
COULD YOU SAVE YOUR TEARS FOR GRADU-ATION?
I'M GOING TO CRY AGAIN AT GRADUATION!
HM?
I'M GETTING DÉJÀ VU.
"I WAS LUCKY ENOUGH TO BE BORN A GIRL!"
"WHU-WHY CAN'D I BE OSGAR?!"
"YOU NEVER KNOW WHAT SOMEONE MIGHT ACCOMPLISH!"

I REMEMBER NOW!
I KNOW YOU HAD A ROUGH TIME AT THE CULTURE FESTIVAL...
BUT FROM HERE ON OUT, YOU GET TO BE THE ADVISOR FOR A BRAND-NEW FIRST-YEAR STUDENT.
YOU CAN'T LET YOURSELF BE A SOBBING MESS IN FRONT OF *HER!*
SO CHIN UP, OKAY?

HEY, GUYS!
WANNA TAKE SOME PICTURES IN OUR CLASSROOM BEFORE THE CEREMONY STARTS?
YEAH!
KA-SNAP
KA-SNAP
KA-SNAP
CAN WE TAKE ONE TO-GETHER?
OUT IN THE HALL?
RISA!
EH?
WHY IN THE HALL?

IT'S A SPECIAL SPOT! THIS IS WHERE YOU TOLD ME YOU HATED ME!
AHA HA!
KA-SNAP
YOU *REALLY* ARE SOMETHING ELSE, HIJIRI.
WITHOUT IT, WE WOULD NOT HAVE THE HONOR OF GRADUATING TODAY.
AND THE LOVE AND SUPPORT OF OUR FAMILIES AND FRIENDS...
AS WE GRADUATE TODAY...
BATHED IN THE WARM LIGHT FROM ONE ANOTHER...
LET US NEVER FORGET THE HOPE, HARD WORK, AND FRIENDSHIP...
THAT GUIDED US THROUGH OUR TWO YEARS HERE.

HOLD TRUE TO OUR TENETS. STRENGTH, RIGHTEOUS-NESS, BEAUTY...
AND WALK WITH YOUR HEAD HELD TALL...
AS MEMBERS OF THE KOUKA THEATER TROUPE!

THAT WAS THE STUDENT REPRESENTATIVE OF THE FORTY GRADUATING MEMBERS OF THE NINETY-NINTH CLASS...
TAKEI TOMOMI!
CLAP
CLAP
CLAP
CLAP
RISA-SENPAI!
WOW, TALK ABOUT AN UGLY CRY.
THEY'RE FILMING US FOR TV, SO TRY TO CRY A LITTLE CUTER, OKAY?
OKAY!
CONGRATULATIONS, HIJIRI-SENPAI.
I *KNEW* YOU WOULDN'T CRY, NARATA-SAN!
WOULD YOU HAVE PREFERRED IF I DID?
I JUST DON'T.

JUST KIDDING.
NO, YOU'RE FINE.
ALL I ASK OF YOU, NARATA-SAN...
IS THAT WHEN YOU GRADUATE AND JOIN THE TROUPE...
YOU TELL EVERY-ONE HOW MUCH I DID FOR YOU.
YOU *DID* DO A LOT FOR US, THOUGH.
WHEN SARASA WANTED TO GO BE WITH HER FAMILY DURING THE CULTURE FESTIVAL...
YOU SAID ALL THOSE HARSH THINGS TO MAKE HER FEEL BETTER ABOUT GOING.
......
YOU'RE GIVING ME TOO MUCH CREDIT.
RISA!!

GREAT JOB! YOU'RE FINALLY A KOUKA GIRL!
WHAP
WHAP
YOU'RE SO BEAUTIFUL!
JUST LIKE YOUR MA!
THE APPLE DIDN'T FALL FAR FROM THE TREE, HUH?
HIJIRI.
MOM.
YOU LOOK LOVELY, HIJIRI. ABSOLUTELY STUNNING.
DAD COULDN'T MAKE IT.
I'M SO, SO PROUD OF YOU, HIJIRI.
YEAH.

HE SENT ME ALL THESE TEXTS...
APOLO-GIZING AND SAYING HOW SORRY HE WAS.
ESPECIALLY AFTER HE MISSED THE CULTURE FESTIVAL, TOO.
I LOVE YOUR FATHER, BUT HE'S KIND OF A COWARD.
OUR NEW HOUSE IS A LITTLE ON THE SMALL SIDE.
COZY, THOUGH! YOU'LL SEE WHEN YOU COME VISIT!
GRADUATES AND FAMILIES!
THE TROUPE JOINING CEREMONY WILL BE HELD IN TWO HOURS!
WE'D LIKE YOU TO GATHER IN THE CLASSROOM TO EXPLAIN WHAT THAT WILL ENTAIL.
YES, MA'AM!

GRADUATES AND FAMILIES...
FIRST OF ALL, CONGRATULATIONS AGAIN.
IN TWO HOURS' TIME, AT THE KOUKA CONCERT HALL...

THIRTY-NINE OF YOU WILL JOIN THE KOUKA THEATER TROUPE.
HUH?
ONLY THIRTY-NINE?
YOUR ASSISTANT CLASS REPRESENTATIVE NOJIMA HIJIRI-SAN...
HAS DECIDED...
SHE WILL *NOT* BE JOINING THE TROUPE.

"I'M NOJIMA HIJIRI. NICE TO MEET YOU, TOO!"
"YOU'D BE WAY BETTER PLAYING THE ROLES FOR OLDER WOMEN AS AN ONNAYAKU!
"THOSE ROLES WOULD BE PERFECT FOR YOU! EVERYONE SAYS SO!"
"BUT NOW I KNOW FOR SURE YOU DON'T LIKE ME.
"THAT'S TOO BAD!"
"I'M SO EXCITED...
"I'M SO, SO EXCITED!
"GOSH, IF ONLY...
"THIS CULTURE FESTIVAL COULD LAST FOREVER!"

HIJIRI!!
WHAT ARE YOU DOING HERE?
DON'T YOU HAVE THE JOINING CEREMONY SOON?
I TOLD EVERYONE I WAS GOING TO COME SAY GOODBYE...
BUT THE TEACHERS SAID THE MEDIA IS HERE, AND COMING TO SEE YOU WOULD DRAW ATTENTION TO YOU...
BUT I--
TRYING TO PRETEND WE'RE FRIENDS? *REALLY*?
I KNOW YOU DON'T LIKE ME, RISA.

I DON'T, BUT I STILL THOUGHT YOU'D BE A TOP STAR!!
YOU'RE A JERK AND THE MOST PASSIVE-AGGRESSIVE PERSON I KNOW...
BUT I *KNEW* YOU HAD WHAT IT TOOK TO STAND FRONT AND CENTER!
SO WHY...
YOU'RE RIGHT.
I *WAS* GOING TO BE A TOP STAR.

I'D HAVE BEEN THE TOP MUSUMEYAKU, BELOVED BY ALL...
SMILING BY THE SIDE OF OUR OTHER TOP STARS.
MY FATHER'S COMPANY WENT UNDER.
I DON'T HAVE THE MONEY TO GO ON LIVING THIS DREAM.
SO...
MY DREAM ENDS HERE.

YOU NEED MONEY TO GET BY DURING YOUR FIRST FEW YEARS, YOU KNOW?
YOU COULD WORK PART-TIME...
ARE YOU *STUPID?*
A KOUKA ACTRESS, WORKING SOME TWO-BIT PART-TIME JOB?
I'M FLABBER-GASTED YOU'D EVEN *SUGGEST* IT.
I KNOW, BUT...
HUH?
SIT.
SIT DOWN! GOD! YOU HAVE TO GO SOON!
SOME-ONE'S GOT TO FIX YOUR HAIR!

HEY.
MY STAGE NAME IS...
ARIOTO SARI.
"ARIOTO" IS THE NAME OF THE BRIGHTEST STAR IN THE BIG DIPPER.
THE JAPANESE NAME, AT LEAST.
STARS SHINE BRIGHT ALL ON THEIR OWN.
AND YOU CAN SEE THE BIG DIPPER MOST OF THE YEAR ROUND.

WOW.
NO ONE'S GOING TO REMEMBER *THAT* TERRIBLE NAME.
OH, SHUT IT!
WELCOME, GRADUATES ...
TO THE WORLD-RENOWNED KOUKA THEATER TROUPE!
KOUKA
Kouka Theater
THESE PAST TWO YEARS, YOU'VE LEARNED THE SKILLS AND TECHNIQUES...
AND HONED THEM FOR THIS VERY MOMENT.
FROM THIS DAY FORWARD, YOU ARE PART OF THE KOUKA LEGACY.

DURING THE REVUE PORTION OF THE AUTUMN TROUPE'S SHOW...
YOU WILL ALL PERFORM A LINE DANCE NUMBER.
THAT WILL BE YOUR DEBUT PERFORMANCE.
AS YOU ALREADY KNOW...
THE KOUKA STAGE HAS A SPECIAL APRON JUTTING OUT AROUND THE ORCHESTRA PIT THAT WE CALL THE **SILVER BRIDGE.**
THE *ONLY* TIME YOU'LL DO A LINE DANCE ON THAT SILVER BRIDGE IS DURING YOUR SPECIAL DEBUT PERFORMANCE.

BUT EVEN AFTER YOUR DEBUT, I'M SURE YOU'LL SHINE AND SPARKLE ON THAT STAGE LIKE THE RADIANT, BEAUTIFUL FLOWERS YOU ALL ARE!
HIJIRI!
I'M HERE AT THE STAGE DOOR FOR YA!!
CAN I GET YOUR AUTOGRAPH?! ☆

YUUTA!!
I...!
I WAS SO *MEAN!* I BECAME A BITCH TO THE OTHER GIRLS!
I'M A HORRIBLE PERSON!
THAT'S WHY KARMA'S PUNISHING ME.
BITCH OR NOT, I'LL ALWAYS LOVE YOU, HIJIRI.
CONGRATS ON GRADUATING.

MY BEST FRIEND IS THE SWEETEST GUY IN THE WHOLE WORLD.
I'M SO LUCKY.
KYAAAA!
KLAK

SO THIS IS THE SILVER BRIDGE.
IT'S ONLY A HUNDRED AND TWENTY CENTIMETERS WIDE!
HEY, RISA!
DID YA HEAR WHY HIJIRI QUIT?

APPARENTLY SHE FELL *HARD* FOR A GUY AND THREW IT ALL AWAY FOR HIM!
TALK ABOUT *DRAMA!*
STRAIGHT OUT OF A KOUKA SHOW!
WOW. OKAY.
THAT'S NOT MUCH OF A REACTION!
THAT'S NOT WHY.
NOT THAT I *KNOW* THE REASON, BUT...
TAKEI'S RIGHT.
AFTER ALL...
I KNEW HIJIRI...

AND SHE LOVED THE KOUKA STAGE MORE THAN ANYTHING ELSE.
ALL RIGHT, LET'S GET THE DRESS REHEARSAL STARTED!
IT'S YOUR FIRST TIME ON STAGE! BREAK A LEG!!
DON'T FALL OFF THE SILVER BRIDGE, MMKAY?
YES, MA'AM!!

THAT GIRL I HATED WILL NEVER CROSS THIS SILVER BRIDGE.
BUT I WILL.
AND I KNOW...
THAT ONE DAY...

Kageki Shojo!! Volume 7 / END

☆ Special Thanks ☆

Asai-san

Momo-san

Nanami-san

Tom-san

☆

Takisawa-san

Kuroki-san

Morita-san

☆

Nono & Jill

☆

♡ all of ♡
my readers

♪☆ヽε(*´◡`*)3人(廿_廿)ﾉ☆♪

Love from

斉木 久美子
Kumiko Saiki

NEXT Kageki Shojo!!
WITH SARASA GONE HOME...
I WILL STAND IN FOR HER.
AI TAKES THE STAGE!
ACTING IS ABOUT TEAM-WORK!
HOW WILL THE FIRST-YEARS REACT?!
WE'RE THE 100TH CLASS OF THE KOUKA SCHOOL!
What will happen at the Culture Festival?!
STRONG!
BEAUTI-FUL!
RIGH-TEOUS!
KAGEKI SHOJO!! 8
COMING SOON!!
Presented by KUMIKO SAIKI